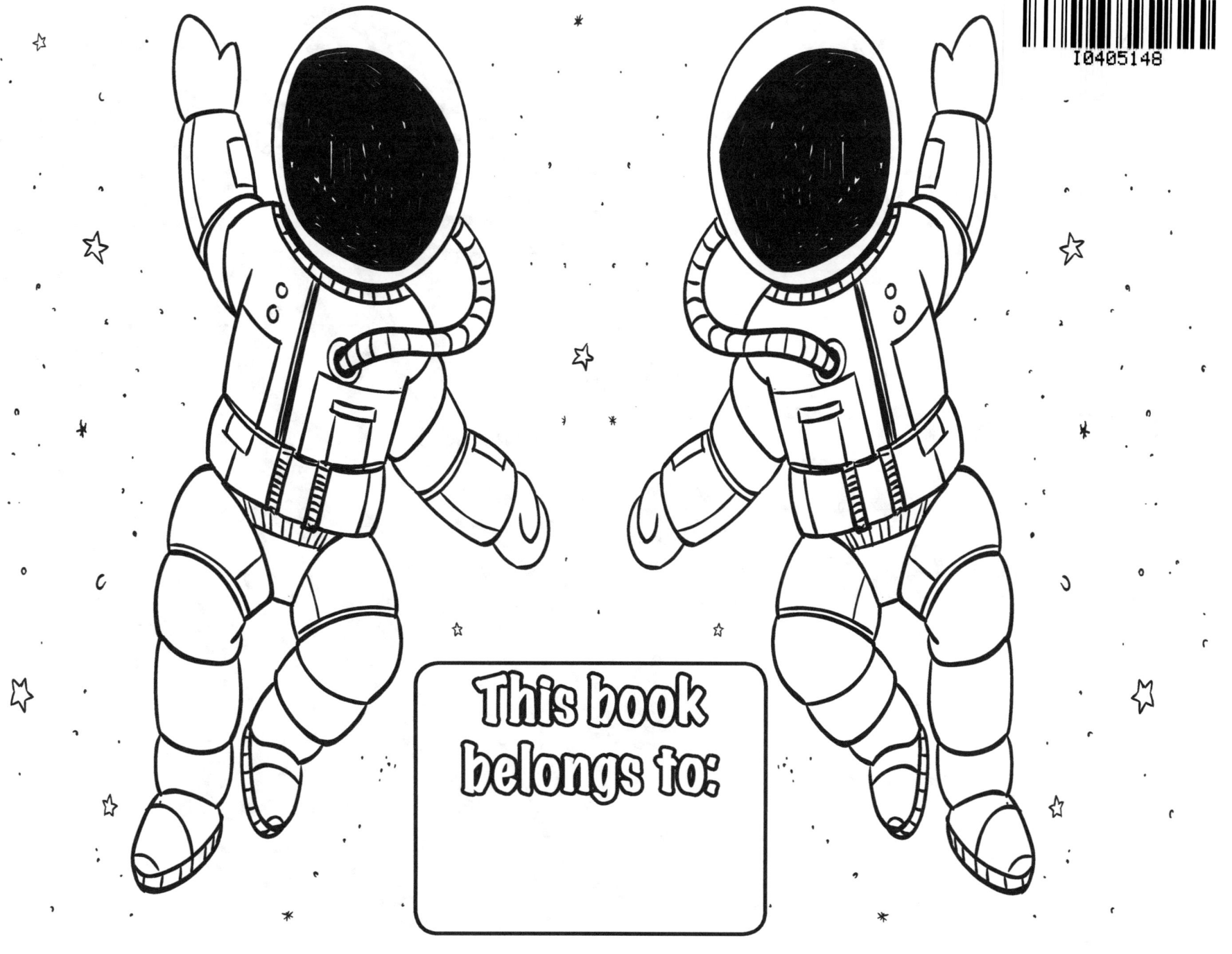

Color With Me!

Mommy or Daddy & Me

SPACE Coloring Book

Mary Lou Brown
Sandy Mahony

Copyright ©2016
Mary Lou Brown
Sandy Mahony

All rights reserved. No part of this book may be reproduced in any form or by any electronic or mechanical means including information storage and retrieval systems, without permission in writing from the authors. The only exception is by a reviewer, who may quote short excerpts in a review.

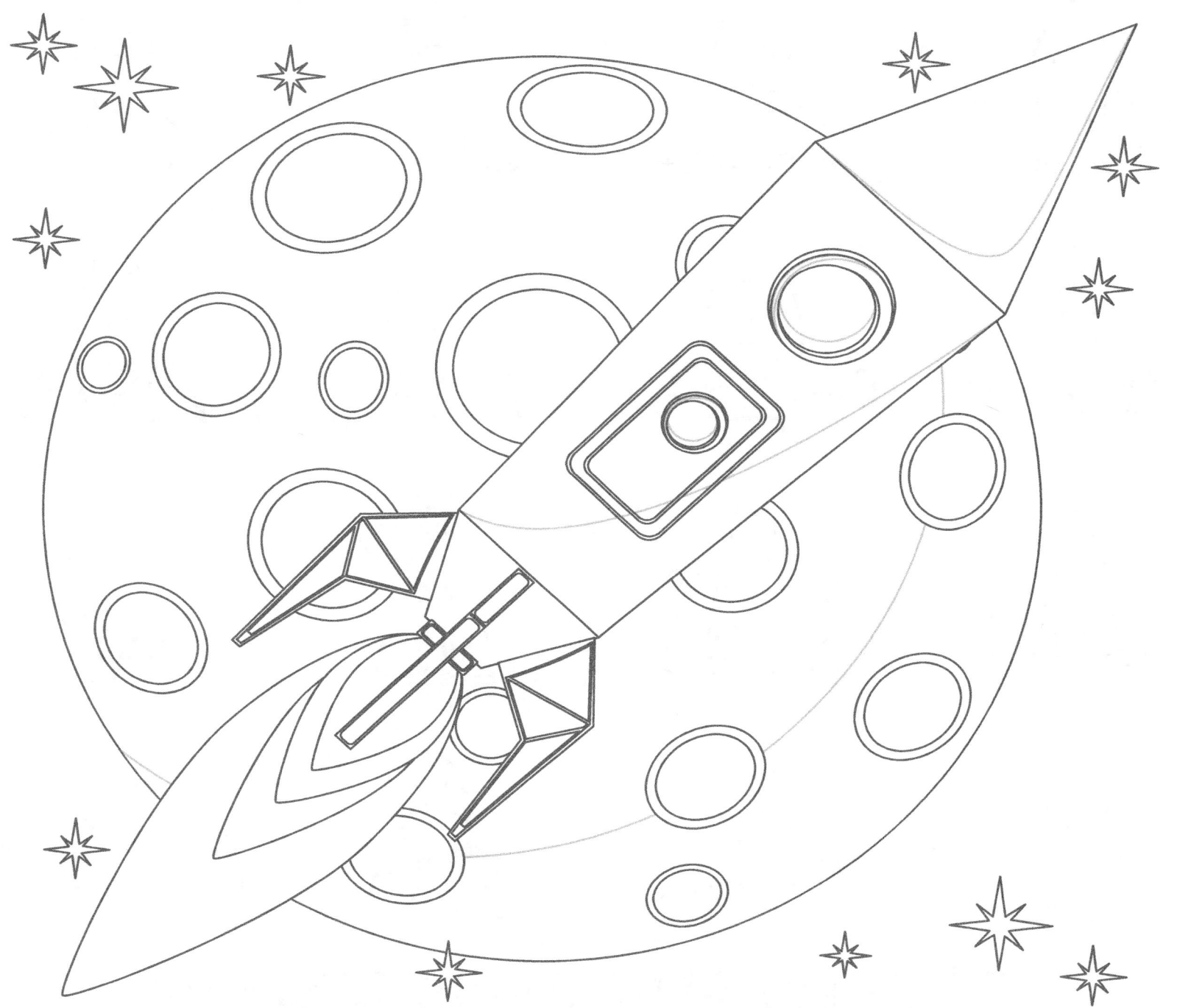

adventurelearningpress.com